Soft Technologies, Hard Choices

Colin Norman

Worldwatch Paper 21
June 1978

This paper is based on the author's research for a book on technology and society, to be published in 1979. Research for the paper was supported by the United Nations Environment Program.

Sections of the paper may be reproduced in magazines and newspapers with acknowledgement to Worldwatch Institute. The views expressed are those of the author and do not necessarily represent those of Worldwatch Institute and its directors, officers, or staff; or of the U.N. Environment Program.

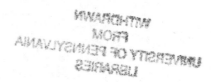

Printed on recycled paper

Table of Contents

I n the British Parliamentary election campaign of 1964, Labour Party leader Harold Wilson repeatedly promised to lead the nation to prosperity through the "white heat of technological revolution."Across the Atlantic, the Apollo Program was in full wing, with the goal of putting a man on the moon by the end of the decade. In Asia and Latin America, the Green Revolution was about o get under way, and it seemed to promise a technological solution o the world's food problem. As for energy, although it was generally cknowledged that oil and gas reserves would not last forever, uclear power was waiting in the wings, ready to provide electricity 'too cheap to meter."

5

Technology seemed to hold the key to many treasures in the fifties nd sixties, as a stream of technological innovations changed virtual-y every facet of life in the industrial world. The postwar recovery of Europe and Japan heralded a period of rapid economic growth, nd foreign aid programs were constructed around large, high-tech-nology projects in the expectation that economic development would quickly follow. But by the early seventies, some of the bright hopes nvested in the technological revolution began to dim.

In the industrial countries, rising concern about environmental pollu-tion, the Vietnam war, and the consequences of unlimited growth in material consumption focused attention on the negative side effects of some technological developments. In the Third World, it has be-come clear that the technological revolution has bypassed most of the world's poor. Although the Green Revolution has increased grain yields, for example, chronic malnutrition is still a fact of life and death for at least half a billion people—a grim reminder that technical fixes cannot solve complex social problems.

The transition from the gung ho technological optimism of the post-war era to the more uncertain mood of the seventies is symbolized by two events of singular technological importance—the 1969 moon landing and the 1973 Arab oil embargo. The moon landing marked the pinnacle of a long, spectacular effort that demonstrated human-ity's technological prowess. Just four years later, however, the oil

The author wishes to thank Daniel Greenberg, Sandy Grimwade, Nicholas Wade, and Charles Weiss for reviewing the manuscript.

embargo underlined the fragility of the industrial world's petroleum-based economies, and provided a forceful demonstration of the close bonds between modern technology and a finite, shrinking resource.[1]

The persistence of deep-rooted social, economic, and environmental problems in rich and poor countries alike has provided fertile ground for questioning the nature and direction of technological development. A few have rejected the values of modern technological society, while others have espoused the "small is beautiful" philosophy put forward by E. F. Schumacher.[2] In general, the prevailing attitude has changed from confidence that technology will pave the way to a better future, to uncertainty, summed up in the query "If we can put a man on the moon, why can't we. . .?"

The uncertainty is understandable, for the world faces an uncertain future. Economists are unable to diagnose, let alone cure, the economic ills that have afflicted most countries in the seventies. Unemployment has reached epidemic proportions in much of the developing world, and it shows every sign of rising in the next few decades. In the industrial countries as well, joblessness is at unacceptable levels. Income gaps between rich and poor countries, and between rich and poor within many countries, have been widening in recent years, a trend that is raising justified demands for greater global equity. The longevity of the world's oil and gas reserves is in doubt, and rich and poor countries alike face the necessity of switching to new sources of energy supply in the next few decades. And there are signs that pressures on many of the world's ecosystems are reaching unsustainable levels.

Those four concerns—employment, equity, energy, and ecology—are likely to remain high on the international agenda for the remainder of the twentieth century and beyond. They must be taken into account in the choice of technologies within countries and in the transfer of technologies between countries. These criteria have not played a prominent role in technological development during most of the postwar era, however.

During the fifties and sixties, unemployment was relatively low in the industrial countries, capital was abundant, energy was cheap and seemingly boundless, and raw materials were available in copious quantities. Technological development therefore generally led to the substitution of capital and energy for labor in the production of goods and services. Technologies have become more complex, energy-intensive, labor-saving, and larger in scale, and industrial society has acquired a voracious appetite for raw materials. Those trends must be examined in the light of the changing global environmental and economic prospects.

Those prospects require that technologies be adapted to the use of constrained rather than abundant resources. But no technology—however appropriate—will solve social problems by itself. The development, introduction, and international transfer of technologies involve a constellation of government policies, vested interests, and political and economic trade-offs. Those factors all constrain the choice of technologies. Moreover, an attack on problems of poverty, malnutrition, disease, and land degradation requires political will, as well as material resources, to overhaul credit facilities, mount adequate public health programs, and institute land reforms.[3]

Nevertheless, the choice of inappropriate technologies can only exacerbate social, economic, and environmental problems. It is clearly time to shed the notion that the biggest, fastest, most modern technologies are always the best, and to seek alternatives that are more compatible with the changing global conditions of the final quarter of the twentieth century.

Employment Impacts

Unemployment on an unprecedented scale has emerged as one of the most pressing political and social problems of the seventies. While governments in industrial countries have been grappling with a pernicious combination of inflation and unemployment, rates of joblessness throughout the Third World have reached extraordinary levels. Two ominous features of the global employment picture stand out: the job shortage will probably worsen before it improves, and it is unlikely that conventional economic remedies will offer sufficient relief.

Economic development theories that held sway during the fifties and sixties are beginning to lose their credibility in the light of mounting unemployment and underemployment throughout the Third World. And as this occurs, it is becoming evident that a massive transfer of modern technology from rich to poor countries will not provide the key to prosperity in the developing world. Meanwhile, in the industrial countries the link between technology and jobs—particularly in the area of energy policy—is being re-examined in the face of unemployment levels unmatched since the Great Depression.

In part a result of population growth during the past few decades, massive unemployment in the Third World has been a long time in the making. It will take even longer to abate. About 200 million people have flooded the labor markets of developing countries during the seventies, and an additional 700 million are expected to require employment by the turn of the century. Already, the number of prospective workers has greatly outstripped the supply of new jobs. By the mid-seventies nearly 300 million people, more than three times the number who have jobs in the United States, were believed to be unemployed or severely underemployed, eking out a precarious existence as casual laborers, street peddlers, shoe-shine boys, and other fringe workers.[4]

More than 30 million jobs must be created each year over the next 20 years merely to keep pace with expansion in the Third World's labor force. Anything less is likely to aggravate inequities and lead to rising levels of poverty. If at the same time productive employment is provided for those who are now grossly underemployed—a critical dimension of any effort to lift the incomes of the poorest people—about one billion new jobs must be created by the year 2000.

These figures provide a central reason why modern technologies cannot be a panacea for development: the capital needed to create enough jobs in modern industries and in Western-style agriculture would be staggering. It now costs an average of $20,000 to establish a single workplace in the United States, and modern industrial jobs in the Third World are no cheaper to create. It requires only a pencil and the back of an envelope to demonstrate the difficulty of establishing a billion jobs at those prices, to say nothing of the energy and materials that such a task would require.[5]

Indiscriminate transfer of modern technology from industrial countries to the Third World can cause more problems than it solves. Technological development since the Industrial Revolution has led to the substitution of capital and energy for human labor in the production of goods and services, substitutions that generally reflected the relative availability and cost of capital, energy, and labor in the industrial world. But these capital-intensive, energy-consuming, labor-saving

"There can be no universal blueprint for
an appropriate technology for any
particular task."

technologies make lavish use of the very resources that are scarce and expensive in the Third World, while failing to utilize much of the Third World's most abundant asset—people.

In general, technologies are economically efficient if the factors of production—labor, capital, energy, and raw materials—are blended together roughly in proportion to their cost and availability. The guiding economic principle should be to maximize the output of the scarcest factor. Since the availability and cost of these four factors vary between rich and poor countries, it follows that different countries require different technologies—or at least different mixes of technology—to make the best use of their resources. There can therefore be no universal blueprint for an appropriate technology for any particular task.[6]

9

Nevertheless, for the developing world in general, technologies that use locally-available raw materials, serve local needs, and can be maintained without sophisticated repair and maintenance services will usually be cheaper to develop and operate than imported technologies. Moreover, by stimulating local innovation and reinforcing other development efforts, simple technologies can lead to self-sustaining development. Although increasing attention is being paid to the use of such technologies in some developing countries, a recent World Bank report candidly notes that "this potential remains largely unexploited."[7]

Governments do not deliberately plan to have a large portion of their labor force unemployed or underemployed, but often that is precisely the outcome when a poor country invests most of its national savings in imported capital-intensive technology. Such investments do raise the productivity of a few workers, and the gross national product consequently increases. But this approach leaves little capital to aid small farmers, landless laborers, and small-scale manufacturers—producers who now constitute the majority of the labor force in most developing countries.

Many developing countries have sought to implant U.S.-style agriculture in their fields by subsidizing imports of heavy machinery and labor-saving techniques, often with assistance from international lending institutions. Pakistan, for example, received a loan from the World Bank in the late sixties to buy some 18,000 large tractors. A subsequent Bank study, which has generated considerable controversy, provides a sobering warning of the danger in assuming that technologies appropriate in one country will confer the same benefits in another setting. Farmers who bought tractors found it easier to work larger farms so they increased their holdings by displacing tenants and by buying extra land. On the average, farm sizes doubled after the introduction of tractors, while labor use per acre dropped by about 40 percent. Yet yields per acre showed little change. The Bank's report concluded that "the widespread introduction of tractors in Pakistan agriculture in the future, if it followed the course that was manifested in the past, would be little short of a disaster to the economic and social fabric of the rural sector."[8]

Such an experience should not lead to the blanket conclusion that *all* capital-intensive modern technologies are inappropriate in the developing world. Far from it. Often, there may be no feasible alternative to sophisticated technologies developed in the industrial world. Imported modern technologies may offer significant advantages in the production of certain goods, such as chemical fertilizers, that are essential for development. And a country that seeks to earn foreign exchange by exporting manufactured goods to the industrial world may be forced to use capital-intensive technologies to bulk-produce high quality merchandise that can compete on the international market. Nevertheless, faced with chronic shortages of capital and rapidly swelling labor forces, most poor countries need to find productive employment for large numbers of people with small expenditures per worker.

As most of the population in developing countries now lives in the countryside, most of the increase in the labor force will also come from the rural areas. If the crushing urban migration that has taken place during the past few decades is to be halted, productive employment must be created in the fields, villages, and small towns. All the above considerations point to the need for technologies that will create employment for landless laborers, lead to more productive use of labor in public works programs, and establish labor-intensive industries.

Demand for rural labor in developing countries fluctuates according to the season. During planting, weeding, and harvesting, every available person is usually busy in the fields from dawn to dusk, but at other times of the year jobs are scarce. A shift to more intensive cultivation can greatly increase agricultural employment, but in regions

10

"If crushing urban migration is to be
halted, productive employment must be
created in the fields, villages,
and small towns."

where the growing season is short, new technologies may be required to allow more than one crop to be grown each year. Selective mechanization, for example, may be needed to speed up planting and harvesting to squeeze additional crops into the growing season. A variety of inexpensive pedal-operated machines, designed to ease and shorten some operations, have been developed in recent years. The International Rice Research Institute (IRRI) in the Philippines has developed a range of relatively inexpensive power tillers, threshers, and weeders for rice cultivation. And a project financed by the World Bank has introduced draft oxen into cotton-growing areas of the Ivory Coast, a technology transfer that has provided significant improvement over hand cultivation at a fraction of the cost of tractors.[9]

11

Irrigation alone can increase labor demand per acre by up to 80 percent by extending the growing season to permit multiple cropping. But the construction and operation of irrigation systems is often a costly business and wealthy farmers with large holdings are usually the first to benefit from irrigation. There are, however, cheaper alternatives. The use of locally-available bamboo or baked clay as filters, instead of metal screens, can cut the cost of a single well to about $15, and a reliable and easily maintained hand pump has been developed for about $100. Windmills constructed with local materials are providing low-cost irrigation in the Omo Valley in Ethiopia. The World Bank is also experimenting with a scheme in India that involves renting portable diesel pumps to farmers for short periods, a strategy that spreads the capital cost and brings them within the reach of small farmers.[10]

High-yielding varieties of rice can also greatly increase labor requirements, largely because they need additional applications of fertilizers and pesticides. A study in Bangladesh found that labor requirements on unmechanized farms (farms that used oxen rather than tractors for draft power) were increased by between 30 and 50 percent when high-yielding varieties were used. When mechanization was introduced along with the new seeds, however, labor requirements dropped. There was little difference in yields between the mechanized and unmechanized farms, which implies that capital invested in mechanization does not necessarily generate more output per acre.[11]

Public works programs, such as the construction of dams, irrigation canals, roads, and buildings, consume a large portion of the budgets of developing countries. Such projects are of two types. Some employ the same technologies that are used in the industrial countries—bulldozers, mechanical diggers, tar spreaders, and so on—and they are consequently highly capital-intensive. Others employ armies of people to move earth with head baskets and shovels. These projects create jobs, but they involve heavy toil and take a long time.

Studies by the World Bank and the International Labour Office have indicated, however, that there is considerable scope for using more efficient labor-intensive methods. Improved wheelbarrows, ox-carts, and hand-operated rail carts to haul materials, ox-drawn plows to break up the ground, and block-and-tackle systems to help move heavy loads can all reduce back-breaking toil and raise productivity to the point where labor-intensive construction is cheaper than capital-intensive methods. Moreover, if the improved tools are fashioned locally with widely available materials, employment is also created indirectly. This may require that local industries be upgraded.[12]

The Chinese have made extensive use of such public works programs. The most famous example is Tachai, where small rocky fields that were frequently washed out by heavy rains have been transformed by hand into productive units through the construction of drainage tunnels, terracing, and the replacement of lost topsoil with earth carried down the mountainside. Similarly, in Lin County, a 1500-kilometer canal network was dug with manual labor during the sixties to irrigate arid and relatively unproductive fields. Such projects have not only absorbed slack agricultural labor, but also greatly increased the productivity of the land, raising demand for agricultural workers. In a sense, the Chinese have harnessed reserves of idle labor and used them for capital formation—a manifestation of Marx's description of capital as congealed labor time.[13]

The use of labor-intensive construction technologies can be limited, however, by shortages of organizational skills. Since a few machines are easier to organize than a large number of workers, there is often a strong incentive to use capital-intensive technologies in place of people.

Manufacturing technologies developed in industrial countries, like agricultural and construction technologies, are often ill-suited to the needs of the Third World. Not only do they require large amounts of capital and provide few jobs, but they often use materials that are not available locally, produce large volumes of goods for remote markets, and need sophisticated repair and maintenance services. It has been assumed that large-scale modern industries would be efficient in developing countries because they take advantage of economies of scale. But such hopes have often proven false. Factories are frequently operated at less than full capacity, which means that capital invested in the plants is used inefficiently and employment is kept well below its potential. Large-scale, centralized production also requires dependable, cheap transportation for the supply of raw materials and the distribution of finished products, but in many developing countries, transportation facilities are inefficient and expensive.

Mounting evidence of such problems has begun to focus attention on the role of labor-intensive, small-scale industries in providing employment and promoting development. In many developing countries, small-scale enterprises, ranging from village artisans to textile producers, constitute the bulk of manufacturing employment. Such enterprises are often shoestring operations, however, lacking access to capital and established markets.

Deliberate attempts to foster small-scale industries, instead of replacing them with large-scale production technologies, have produced encouraging results in a few countries. China's rural industries are perhaps the best known. According to one estimate, there may be as many as 500,000 rural industrial units in China, producing items such as cement, fertilizer, iron and steel, agricultural machinery, textiles, and processed food. They rely for the most part on local materials and supply local needs. Like China's rural public works programs, the rural industries are geared toward improving agricultural productivity, a process that creates employment directly in the factories and indirectly in the fields.[14]

Although there has been considerable debate about the efficiency of China's small-scale industries, a team of American experts who visited China in 1975 under the auspices of the National Academy of Sciences generally found them to be effective in stimulating rural development. The failure of many "backyard" iron and steel plants established during China's "Great Leap Forward" in the late fifties and early sixties, however, shows there are limits to the extent that some plants can be scaled down and remain economically viable.[15]

14

Labor-intensive, small-scale industries play a key role in the development policies of South Korea and Taiwan as well. They form links between agriculture and manufacturing and provide some inputs into the modern, large-scale industries that have been established in those countries. By decentralizing its industries and ensuring early integration between the farming and industrial sectors, Taiwan, like mainland China, has managed to curb migration from the countryside.[16]

India has also experimented extensively with small-scale, labor-intensive manufacturing. When Gandhi led the Indian people to independence from Britain, his vision—summed up by the choice of the spinning wheel as the symbol of the independence movement—was of decentralized "production by the masses." During the fifties and sixties, however, the Indian government invested heavily in large urban-based industries, and Gandhi's concept of village and cottage industries took a back seat.[17] But rising unemployment and underemployment in India, coupled with widespread flight from the land, has refocused attention on the potential for decentralized industries to provide productive, low-cost jobs in the countryside. The ruling Janata (People's) Party pledged in November 1977 to dismantle urban-based textile, shoemaking, and soapmaking industries and to move their production to the villages.[18]

Domestic research and development can sometimes produce a more appropriate alternative to imported manufacturing technologies. A good example is the development of small-scale sugar plants in India. In the fifties and sixties, several modern factories were established in India to produce white sugar from locally-grown cane, but farmers in remote areas were not able to sell their sugar to the plants and the processing capacity proved insufficient to use all the cane that was available. The Planning Research and Action Institute in Uttar Pradesh developed an alternative technology suitable for small plants serving local markets. The comparison between the two technologies is striking: an investment of 28 million Rupees can establish one large plant capable of producing about 12,000 tons of sugar a year with 900 employees; the same investment can build 47 small plants with an output of about 30,000 tons and a total employment of nearly 10,000.[19]

"Small producers who lack
financial resources are in no position
to experiment with unproven
technologies."

As these examples indicate, low-cost technologies designed to increase employment are finding growing use in some countries. But the difficulties in taking an alternative to the high-technology route to development should not be underestimated.

All technologies require extensive development and testing before they can be widely used. Low-cost technologies are no exception. Indeed, considerable ingenuity is often required to scale down production processes and to develop equipment that can be easily maintained by local people. Small producers who lack financial resources are in no position to experiment with unproven technologies.

Notions of prestige can influence governments to choose the latest modern technology when more appropriate ones are available. Many developing countries lack domestic research and development programs and have little capacity to innovate. Moreover, when capital is underpriced by such factors as government-subsidized credit arrangements, tax breaks, and overvalued exchange rates—while labor costs are elevated by powerful unions operating in the modern industrial sector—there is a strong incentive to choose capital-intensive, labor-saving technologies.[20]

Most important, without social and political changes that redistribute income, overhaul inequitable land ownership patterns, reform credit systems, and provide support for small farmers and manufacturers, appropriate technologies will be difficult to introduce. Powerful vested interests support large-scale manufacturing, mechanized farming, and other symbols of modernity.

While developing countries are facing the certainty of rising unemployment in the coming decades unless present trends are reversed, the outlook for the industrial world is more difficult to gauge. Whatever the future holds, present unemployment levels in the industrial world allow no basis for complacency. Seventeen million workers were idle in Northern Europe, North America, and Japan in 1975. Unemployment reached 8.5 percent in Canada in December 1977; in November, the number of unemployed in the nine countries of the European Economic Community reached six million for the first time in history.[21] In the United States, though total unemployment dropped slightly in early 1978, it still stood at a level that would have been considered intolerable just a few years prior, with inner-city joblessness among black youths at a numbing 40 percent.[22]

15

Energy problems loom large in the unemployment picture in these countries, and they will continue to be a dominant factor. Shortages of natural gas during the 1976/77 winter were directly responsible for idling some 1.8 million U.S. workers, and the economic malaise that has led to current high unemployment rates throughout the Western world can be traced in part to rising energy prices. The choice of energy technologies has both direct and indirect impacts on employment.[23]

16

An energy strategy featuring large-scale coal and nuclear generating plants will require vast amounts of capital expenditure, but will provide few jobs directly. The energy policy along these lines outlined by the Ford Administration in the United States in 1975, for example, was expected to require about $1 trillion in capital expenditure by 1985, an amount that would soak up about 75 percent of all net private domestic investment, compared with about 25 percent in recent years.[24] Such a program would divert spending from other, more labor-intensive sectors of the economy. Conservation programs, on the other hand, generally provide large numbers of jobs for relatively small monetary outlays, and several studies have shown that solar energy technologies are particularly labor-intensive.

A projection of the employment impact of an aggressive solar energy program in California indicated that some 377,000 jobs a year could be created in the eighties. That level of job creation would be sufficient to halve California's present unemployment total. Another study found that while construction and operation of California's controversial Sundesert nuclear plant would provide about 36,300 jobs directly and indirectly, a solar program producing an equivalent amount of energy could create about 241,000 jobs. Solar technologies, moreover, create jobs in the areas where people live, while construction of giant power plants requires work crews to be gathered in one location, disrupting the life of local communities.[25]

With unemployment rates at their present levels and prospects for a return to the surging economic growth rates of the fifties and sixties remote, governments in industrial countries—like their counterparts in the developing world—must pay increased attention to the link between employment and the choice of technologies.

Technology and Equity

In the quarter-century following World War II, production of goods and services tripled in value, and the global economy seemed set on an upward path. In those heady days, government leaders everywhere were able to promise their poorest citizens that economic growth would produce benefits for all. If the economic pie continued to expand, the argument went, everybody's slice would eventually grow larger.

Those promises have been fulfilled for very few. The gap between rich and poor countries has widened into a gulf, and in many nations the wealthy have prospered while the impoverished majority have become relatively worse off. More than a billion people are now thought to be living in conditions of extreme poverty, at least half of them unable to meet their basic food, health care, shelter, and education needs. The coexistence of rising wealth with widespread poverty is not found in every country, nor is it limited to the Third World. But it is clear that rapid economic growth, measured in terms of gross national product, is not sufficient to guarantee a better life for all.

For too long, planners have been preoccupied with growth itself rather than with the nature of growth and the distribution of benefits. Pressure for the redistribution of global wealth, both within and among countries, has been rising during the last ten years. Moreover, since the global economy ran out of steam in the mid-seventies and economic growth is unlikely to resume its earlier pace, the world faces the difficult task of dividing a pie that is expanding more slowly.

The reasons for the concentration of wealth in a few hands are manifold and complex. But one contribution to the forces that preserve and aggravate inequities is the nature of the technologies adopted in many countries. Investments in technology for agriculture, industry, health care, transportation, and energy often benefit only a fraction of the population, raising the living standards of a few and stretching the gap between rich and poor.

The unequal impact of technological change stems from a simple relationship. Goods and services are developed largely in response to demand, and demand comes from those who have power in the marketplace and in the halls of government. Skewed income distribution leads to the development and adoption of technologies that meet the demands of the privileged, and those technologies in turn exacerbate inequities because they lie beyond the reach of the poor. A multinational corporation in the United States claims in its advertising that "technology is a continuing response to the needs of life"; unfortunately, the needs of the poor and the demands of the elite are often not synonymous.[26]

The same skewed distribution of technological supply and demand takes place within and among countries. More than 95 percent of the global research and development budget and most of the world's technological capacity is concentrated in the industrial countries, and consequently the bulk of the world's technological effort is devoted to solving problems of the affluent world. For example, the United States alone spends more than $1 billion every year on research into cancer and heart disease—the major killers in the industrial world— while only a tiny fraction of that amount is spent worldwide on research to control schistosomiasis, a debilitating parasitic disease that afflicts some 200 million rural poor in the Third World. There is a vast collection of published research on sewage disposal, but more than 98 percent of it is irrelevant to the needs of poor countries, according to a survey conducted for the World Bank.[27] Most agricultural research is geared toward raising the productivity of large farmers working temperate zone soils; comparatively little attempts to develop technologies for small farmers in the tropics.

Any reduction of inequities in most poor countries requires not only the creation of large numbers of jobs, but also an attempt to raise the productivity of people who have so far been left out of the development process—small farmers, small-scale manufacturers, landless peasants, and others now barely making ends meet. This should include the development of low-cost production technologies that do not depend upon large-scale enterprises, and the overhaul of credit and land tenure systems to allow small producers to take advantage of new technologies. Technologies alone, whatever their scale, cannot eliminate inequities; social and institutional changes are also required.

The Green Revolution provides a closely studied example of how seemingly beneficial technologies can worsen the lot of small producers when the system is biased against them. Although high-yielding seeds work as well on small farms as on large holdings, they re-

"The health budgets of most poor
countries are heavily weighted
toward modern hospitals
that cater to the urban elite."

quire irrigation and increased use of fertilizers and pesticides—all of
which raises the costs of production. In countries such as Taiwan
and Japan, where egalitarian credit systems allow small farmers to
purchase the necessary inputs, high-yielding dwarf rice has not
exacerbated inequities. But in parts of such countries as India and
Pakistan, small farmers often have difficulty raising credit, and they
are less able to take risks with new agricultural techniques. Benefits
from the Green Revolution have therefore accrued mostly to the
wealthier farmers, and income gaps have widened.[28]

19

In industry, as in agriculture, investments in capital-intensive technologies benefit a narrow section of the population while other parts
of the economy are starved of capital. And concentrating investment
in urban-based industries can also aggravate disparities between
urban wage earners and rural peasants. A study of employment and
income distribution in Colombia in 1970, for example, found that the
living standards of about one-third of the total population had probably declined in absolute terms since the thirties, although incomes
of workers in modern industries had advanced considerably. Similarly, the soaring economic growth rates achieved by Brazil in the sixties
and early seventies benefited primarily the top 20 percent of the population; the remainder, mostly rural peasants, were left relatively
worse off by Brazil's "economic miracle."[29]

Although raising the productivity and incomes of those who are now
grossly underemployed will be a major key to securing greater distribution of wealth, the link between technology and social equity is
not limited to production technologies. Striking examples of how the
choice of technology can determine whether the benefits are available
to all sections of the population can be found in health policies.

The most severe health problems of the Third World are malnutrition,
and infectious and parasitic diseases—illnesses associated with
poverty and poor sanitation. Yet the health budgets of most poor
countries are heavily weighted toward modern hospitals that cater
to the urban elite. Typically, 80 percent of the expenditure on health
in developing countries is devoted to hospital care, and less than
20 percent is spent on preventive medicine accessible to the rural and
urban poor. According to a World Bank estimate, some 800 million
people have no access to even minimal health care.[30]

There are many explanations for inequitable health care priorities, including colonial policies that provided curative medicine for expatriate Europeans, medical education systems that are carbon copies of Western medical schools, and hospital services that respond to the demands of the powerful and affluent in developing countries. As British economist Charles Elliott notes:

> The results are bizarre. In the Philippines, a country in which much of the population has no health care beyond that of the helot, is to be found one of the most sophisticated cardiology units in the world. In the Ivory Coast, the Centre Hospital Universitaire has facilities that few hospitals in France can rival. . . . Such facilities are used at half capacity, but preempt the lion's share of the recurrent budget of the ministry of health (over 50 percent in both cases).[31]

To those examples might be added the three open-heart surgery units in Bogota, the running costs of which are sufficient to provide a pint of milk a day to one-quarter of the city's children.[32]

Expenditures on expensive medical technologies that soak up the bulk of health budgets in poor countries reflect a choice of providing high-quality medical care for a few rather than meeting the basic health needs of many. During the past decade, however, several developing countries have begun to refocus their health policies, training cadres of paramedical workers and establishing medical facilities in the villages and poor neighborhoods. China's 1.3 million barefoot doctors, Sri Lanka's paramedics, Tanzania's Ujamaa village health centers, and Cuba's neighborhood health clinics all provide routine medical services at low cost per beneficiary. They also provide preventive services. The results can be dramatic. In Sri Lanka, a country with an annual income of only $130 per person, life expectancy at birth is now approaching that in the United States.[33]

The choice of transportation technologies can also result in skewed distribution of benefits. During the past few decades, many Third World governments have tried to entice foreign automobile companies into building factories in their countries in an effort to establish modern transportation systems. Yet only a small portion of their citizens can afford to buy cars. And the huge investments on roads, repair and servicing facilities, and fuel stations that would be needed

"If energy systems are founded entirely
on large-scale, centralized power plants,
social inequities will inevitably increase."

to support such a system often involve heavy foreign exchange costs. As World Bank economist Mahbub ul Haq notes of his native country, "During the decade 1958-1968, Pakistan imported or domestically assembled private cars worth $300 million while it could spend only $20 million for public buses."[34]

If governments in developing countries were instead to invest transportation funds in bus and rail services, augmented by private bicycles and mopeds, foreign exchange costs would be reduced, domestic manufacture and servicing would be possible, and some air pollution would be avoided. As Lester Brown points out, "If an index of mobility were constructed for a national population, it would undoubtedly show that a system designed along [these] lines . . . would give far more people greater mobility than an automobile-dominated system can provide."[35]

The industrial countries are not immune to distorted investments in transportation. The $4.2 billion spent to build the Concorde represents a subsidy by British and French taxpayers for international travel by a tiny, affluent minority. And since the aircraft are being operated at a loss by state-run airlines, the subsidy is continuing.[36]

Investments in energy systems can have a profound impact on social equity. The construction of large urban-based electrical generating plants has brought much-needed power to the major cities and industrial centers in developing countries, but it has also aggravated differences between urban elites and the bulk of the population in the hinterlands. Professor A. K. N. Reddy has calculated that about 70 percent of the electrical energy consumed in India goes to urban industries, 15 percent goes to other urban consumers, and only about 12 percent is consumed in the villages. Yet about 80 percent of India's population lives in rural areas. Moreover, according to Reddy, electricity is inequitably distributed within the villages themselves: on average, only 15 percent of the households are electrified.[37]

Such unequal distribution of electrical supply is not surprising in view of the immense cost of constructing centralized facilities and building transmission lines to deliver the power. Thus, if energy systems are founded entirely on large-scale, centralized power plants, social inequities will inevitably increase because few developing countries can afford to extend services to the rural areas. Several recent studies have indicated, however, that decentralized energy sources—based on solar collectors, small-scale hydroelectric generators, firewood plantations, and biogas plants—can provide energy for remote villages more cheaply than centralized power stations and national grids can.[38]

Few countries, however, have yet attempted to build such an energy system. One country that has is China. Although China's major cities are powered mostly by coal-fired generating plants, the villages derive much of their energy from renewable sources, chiefly biogas, firewood, and small-scale hydroelectric power. According to official Chinese reports, some 4.3 million biogas units have been constructed in China during the past three years; and in 1975, there were reportedly 60,000 rural, small-scale hydroelectric plants in operation, an eightfold increase over the 1965 total.[39]

But even seemingly "appropriate" technologies can worsen social inequities in some settings. In the energy field, biogas plants are a case in point. The plants produce methane through the fermentation of a mixture of livestock dung and water (sometimes with the addition of human excrement and crop residues), providing gas for cooking and lighting, and solid residues that constitute an excellent fertilizer. The production of these badly-needed commodities from waste products seems like a good bargain, but there are two major drawbacks. First, the plants require manure from at least three cows to produce sufficient gas for a single family, a requirement that restricts their use to relatively wealthy families. And second, in some countries cattle dung is now collected, dried, and used as fuel by all villagers; it is essentially a free good. The introduction of biogas plants places a premium on dung, however, which can eliminate the poorest villagers' chief source of fuel. Larger plants serving an entire community could get around such problems, but their introduction would require the establishment of new cooperative arrangements.[40] The point is not that biogas plants are inherently bad—indeed, they are one of the most promising solutions to Third World energy problems—but that they cannot be introduced thoughtlessly, without new forms of social organization.

Technologies for housing, water supply, education, and a host of other areas can similarly result in skewed distribution of benefits.

The relationship between technology and equity has only recently become a matter of concern to governments and international lending institutions, however, and the complex interactions between technical change and social benefits are but dimly understood. Nevertheless, social equity must become an increasingly important criterion in public investments in the years ahead or the choice of technologies will only widen the gulf between rich and poor.

Energy Considerations

Cheap and abundant fossil fuels have played a central role in most of the key technological developments since the Industrial Revolution. But that role has been particularly evident during the past few decades. The virtual doubling of global agricultural production between 1950 and 1975 relied heavily on the use of energy-intensive chemical fertilizers. Projected shortages of important minerals have been averted by new technologies that have harnessed increasing amounts of energy to mine and process low-grade ores. And spectacular increases in labor productivity—the basis for unprecedented global economic growth in the postwar era—have largely been achieved with mechanization technologies that replaced human and animal muscle power with fossil fuels.

Most of these technological developments took place when a barrel of oil cost 15¢ at the wellhead and less than $2 on the international market. Cheap oil and gas consequently became the lifeblood of modern industrial society, rising from only 16 percent of the world's commercial energy budget just 50 years ago to about two-thirds of it today. In addition to providing an immense subsidy for agricultural and industrial production, cheap oil profoundly shaped physical and social environments during the past generation. It gave birth to the concept of planned obsolescence; provided inexpensive transportation that changed the face of cities, towns, and the countryside; and helped fuel rising expectations of material wealth. With little incentive to husband energy resources, energy-intensive technologies proliferated, and economic planners paid little attention to the efficiency with which energy was used.

When oil and gas were cheap and plentiful, the marriage between petroleum and modern technology was scant cause for concern. Indeed, as technological advances had overcome food and materials shortages in the past, they were being counted on to avert energy shortages in the future. The vehicle for such hopes has long been nuclear power, whose proponents once confidently promised virtually unlimited energy at prices that would undercut even those of oil and gas. But complacency about energy supplies has been rudely dispelled in the seventies. The 1973-74 Arab oil embargo and the fivefold rise in world oil prices during the following four years marked an abrupt transition from a period of abundance to a new era of rising energy costs and uncertain petroleum supplies. Barring another embargo, the global economy is not on the verge of grinding to a halt because of oil shortages, but the long-term prospects for cheap energy are not good.

Global oil and gas production is expected to peak in the nineties, and to decline steadily thereafter. Some projections have even indicated that oil shortages could develop in the early eighties if key reserves become seriously depleted. Nuclear power has run into myriad problems, among which steeply rising costs, the lack of permanent waste disposal facilities, and fears of weapons proliferation loom large. And a massive switch to coal could have unacceptable health and environmental costs, including the alarming possibility of irreversible changes in the earth's climate resulting from a buildup of carbon dioxide in the atmosphere.[41]

With oil shortages projected, and large question marks hanging over nuclear power and coal, there is an urgent need to develop and deploy technologies that make use of renewable energy resources—direct sunlight, running water, winds, and plant materials. Equally urgent is the need to pay close attention to the efficient use of energy. A good example of how cheap energy influenced technological change —and how rising energy prices and potential scarcities of oil and gas may affect future developments—can be found in the food production and distribution systems of advanced industrial countries.

In traditional agricultural systems, where human labor provides the only source of energy for tilling, planting, weeding, and harvesting, the energy contained in the crops must exceed the energy used in their cultivation—if not, agriculture would be unable to sustain traditional farming communities. Thus, slash-and-burn cultivation of corn in parts of Latin America produces about five kilocalories of

"About 1,150 kilocalories of
fossil fuel energy are needed
to ship one pound of vegetables
from California to New York."

food energy for every kilocalorie of energy spent in the fields. Wet
rice agriculture in parts of Asia offers even better energy returns:
each unit of energy invested in cultivation yields between 10 and 15
units of food energy.[42]

25

At the other end of the scale, the highly mechanized farms of the
United States, the Soviet Union, Europe, and Japan rely on fossil
fuels for most of their energy requirements. According to David
Pimentel of Cornell University, the equivalent of 80 gallons of gasoline is now used to raise an acre of corn in the United States, and
every kilocalorie of energy used in the cultivation produces only
about two kilocalories of food energy. Moreover, if the corn is then
fed to cattle in feedlots, the energy balance is tipped the other way
—at least ten units of energy are invested for each energy unit contained in beef.[43]

The availability of cheap energy for the production of fertilizers,
pesticides, and herbicides meant that yields could be raised without
greatly increasing the land in cultivation. Corn yields in the United
States have in fact more than doubled over the past 30 years, largely
because of a 16-fold increase in fertilizer use. Cheap fossil fuels replaced human and animal power in industrial countries through the
use of tractors, combine harvesters, electric pumps, and other agricultural machines. As a result, the proportion of the population employed in agriculture in the United States dropped by half between
1920 and 1950, halved again by 1962, and since then has dropped
yet again by 50 percent. Only about 4 percent of the American labor
force is now employed directly in the fields.[44]

In a society where only a tiny fraction of the population remains on
the land, vast amounts of energy are required for food storage and
distribution. In the United States, about four times as much energy
is used to transport, process, store, sell, and cook food as to produce
it. Production of some foods has become concentrated in specific
regions. Commercial vegetable production, for example, is now concentrated in California and Florida, although only a few years ago
New Jersey supplied a substantial portion of the vegetables marketed
on the East Coast. About 1,150 kilocalories of fossil fuel energy are
needed to ship one pound of vegetables from California to New York.
That is more energy than most vegetables contain.[45]

Such energy accounting does not imply that the world must revert to traditional agricultural practices because of potential fossil fuel shortages. But it does suggest that Western-style food systems have become excessively dependent on fossil fuels, and that they are not suitable models for most developing countries to follow. If the world's population were fed an American diet, produced with U.S.-style food technologies, production and distribution alone would use up all known global oil and gas reserves in just 13 years. Similarly, if India were to convert its agricultural system to American methods, agriculture alone would require fully 70 percent of the commercial energy that is now used for all purposes in that country.[46]

The energy balance of Western food systems suggests that the chief inefficiencies occur after the food leaves the farm gate. Perhaps the most egregious inefficiency is the use of a two-ton automobile to carry a ten-pound bag of groceries from the supermarket. Excessive processing, transporting, and packaging uses vast amounts of energy that might be saved by less centralized food production and distribution. Urban gardening, solar-heated greenhouses in urban and suburban areas, and greater use of refillable containers in grocery shops would all help make the food system in developed countries less energy-intensive.

Although changes in processing and distribution offer the best prospects for improvements in the energy efficiency of the food systems of most countries, there is considerable scope for improvement on the farm as well. Energy used to produce chemical fertilizers is now the chief energy input into American agriculture, accounting for more fossil fuel than the gasoline used to power tractors and other farm machines. Chemical fertilizers have taken over almost completely from the manures and leguminous plants that were used as fertilizers just a few decades ago, a substitution made economically attractive by cheap oil and gas. As energy prices rise, however, the economic balance may again favor more extensive use of traditional fertilizers.[47] The use of sewage sludge could also reduce requirements for chemical fertilizers, provided serious questions about the contamination of sludge with heavy metals can be resolved. In addition to saving energy, more extensive use of organic fertilizers like manure would help improve the condition of seriously depleted soils, while recycling sewage sludge on farmlands would reduce the capital and energy costs of constructing new sewage treatment plants.[48]

Not all traditional technologies are energy-efficient. The technology of cooking over an open fire—common in many Third World villages —evolved when firewood was abundant and supplies seemingly inexhaustible. But like the use of oil in the industrial world, current patterns of firewood use cannot be sustained indefinitely. In many parts of the Third World, the countryside has been stripped bare of trees, creating a severe energy crisis for hundreds of millions of people, driving up the price of firewood, and causing serious soil erosion problems. Reducing the demand for firewood by improving the efficiency with which fuel is used in cooking could therefore bring benefits even beyond those of saving energy.[49]

Cooking over an open fire requires more energy than cooking over a gas or electric stove because about 90 percent of the heat is wasted. According to energy analyst Arjun Makhijani, the efficiency of traditional Indian cooking stoves could be doubled for an investment of about $10 a stove, saving a family $10-$25 a year in firewood costs. Similarly, a mud stove has been developed in Guatemala that requires only half the wood used in cooking over an open fire. The stove, which costs less than $10, also warms the house and heats water with waste heat. It produces a significant health benefit as well: it replaces the traditional practice of lighting a fire in the middle of a chimneyless room, a practice that promotes lung and eye diseases.[50]

In many developing countries there is substantial room for improvement in the efficiency with which draft animal power is used. Draft animals, the chief power source for cultivation throughout Asia, provide 12 times as much power in India as do tractors, for example. They convert renewable energy sources—roughage from grasslands —into useful work. But their efficiency is often impaired on two counts. First, there have been few improvements in the design of animal-drawn plows, carts, and other implements, or in the livestock breeds, for several centuries. Yet recent studies have shown that relatively simple design changes, such as more efficient harnesses and rubber-wheeled carts, can substantially improve the effectiveness with which animal power is transformed into useful work. And second, cattle dung is often burned for cooking fuel, a practice that deprives the soil of much needed fertilizer and reduces the efficiency of the forage-animal-crop energy cycle. If dung is fermented in a biogas plant, enabling the energy to be extracted and the nutrients to be returned to the soil, the economic contribution of animal power would be greatly enhanced.[51]

The growing energy-intensity of food production systems in both industrial and developing countries has been matched in the transportation sector. The energy used in transporting passengers and freight depends critically on the transportation system used. In most industrial countries, the least energy-efficient ways of moving people and goods have greatly increased their share of the traffic in the past few decades. And the energy efficiency of airplanes, automobiles, and trucks actually declined during the fifties and sixties.

Eric Hirst, an energy analyst from Oak Ridge National Laboratory, has calculated that the energy efficiency of aircraft dropped by nearly half between 1950 and 1970, as jet engines replaced more frugal piston engines. The efficiency of automobiles declined by about 12 percent over the same period, while that of trucks dropped marginally. Yet the airlines' share of freight and passenger markets in the United States rose by factors of 7 and 5 respectively in the fifties and sixties. The number of automobiles climbed from 40 million to 92 million, and trucks increased their share of freight transportation from 13 to 19 percent.[52]

Meanwhile, railroads were more than five times as energy-efficient in 1970 as they were in 1950, largely because diesel engines replaced steam locomotives. Yet the railroads lost freight business to trucks and airplanes; they accounted for 35 percent of the market in 1970 compared with 47 percent in 1950. Their share of passenger transportation dropped from 7 percent to 1 percent over the same period. Similar trends took place in urban transportation as automobiles, the least energy-efficient means of transportation, proliferated and as buses lost passengers.[53]

These shifts toward more energy-intensive transportation technologies have not been accidental. Jet aircraft offer advantages of speed, automobiles provide convenience, and trucks are more versatile than railroad cars. But government policies have also subsidized these less efficient systems through vast highway construction programs, airport building projects, and freight charges that have favored road transport over rail.

"Sales of bicycles in the United States
have outnumbered those
of cars since 1972."

When transportation systems are constructed around the automobile, it is often difficult for other, more efficient technologies to compete. Such is the case with the bicycle in most cities. By far the most energy-efficient means of transportation, the bicycle plays a key role in many developing countries, but in the industrial world it has largely been pushed aside by the automobile. Many Third World cities are also becoming auto-centered. Recently, however, the bicycle has been staging a comeback.

29

Sales of bicycles in the United States have outnumbered those of cars since 1972; the number of bicycles in use in Japan climbed from 10 million in 1950 to 47 million in 1977, and they are now being produced at the rate of more than 6 million a year; and in Britain, yearly sales rose from 467,000 in 1970 to 1.2 million in 1975. One city that has made a major effort to accommodate bicycles in urban transportation plans is Davis, California, where an extensive network of bicycle paths has been constructed. A survey of traffic in Davis in the summer of 1977 found that bicycles represented 40 percent of all vehicles on one heavily traveled street. Urban planners in Dodoma, Tanzania adopted a similar philosophy: the master plan for the city decrees that the ratio of bicycles to cars will be 70:30.[44]

When oil cost $2 a barrel, energy efficiency was not a major preoccupation of transportation planners, but as its price rises, it will inevitably assume a more important role. For developing countries, the rising energy intensity of transportation systems built around automobiles and trucks should provide a warning of the costs involved in following in the footsteps of the industrial world. In the industrial countries, subsidies for energy-intensive transportation systems should be overhauled, and wider use of more efficient ways to move goods and people should be encouraged.

The availability of cheap and abundant fossil fuels has shaped technological developments in many areas other than agriculture and transportation. Advances in building technology, for example, have produced glass-walled, hermetically sealed structures whose chief concession to the external environment is found in the size of their heating and cooling systems. A study by the American Institute of Architects has indicated that improvements in the design of new buildings and modifications to existing ones could save a staggering 12.5 million barrels of oil a day by 1990. Thrifty building design includes adequate insulation, passive solar heating and cooling design, a minimum of lighting, and construction materials requiring less energy in their production, such as stainless steel rather than aluminum.[55]

30 The concept of planned obsolescence also matured in an era of cheap energy. The massive one-way flow of materials through most industrial economies, from mines to garbage dumps, requires vast amounts of energy at every stage of the journey. The extraction and processing of materials alone accounts for about 25 percent of all energy used in the United States, and most of it ends up on the trash heap. The average American generates about 1,300 pounds of solid waste a year, less than 7 percent of which is recycled.[56] Yet the production of steel from scrap requires only 14 percent of the energy needed to produce it from virgin ore; the equivalent figure for copper is 9 percent, and for aluminum, 5 percent.[57] Recycling can never be perfect, however. Reducing the materials consumed by industrial society requires designing products that are more durable and easier to repair, and eliminating wasteful packaging.

Finally, a less obvious source of inefficiency is the use of high-quality energy sources, such as electricity, to perform tasks where energy of a lower quality would be adequate. Electricity is a versatile energy source, capable of performing tasks ranging from the production of high temperatures to powering appliances and subway systems. But roughly three units of fossil fuels are required to generate one unit of electricity—the excess energy is usually rejected into the atmosphere as low-temperature heat. As physicist Amory Lovins notes, "This electricity can do more difficult kinds of work than can the original fuel, but unless this extra quality and versatility are used to advantage, this costly process of upgrading the fuel—and losing two-thirds of it—is all for naught."[58] Using electricity to heat homes and offices to 68°F is precisely the kind of application that does not make use of its quality and versatility.

As the world moves from an era of low-cost, abundant energy to an era when energy costs are bound to rise and oil and gas are expected to become scarce, the technological developments of the past provide neither sound models for the future nor a sound basis for the choice of energy technologies in developing countries.

Ecologically Sustainable Technologies

Two themes have dominated technological evolution for thousands of years: wars, and the struggle by humanity to control nature. In view of the formidable power of modern technology to manipulate biological systems for the production of food and fiber, to combat disease, and to provide protection from some of the vagaries of nature, it is sometimes tempting to conclude that one historic struggle is close to being resolved, at least in the industrial countries. But in the past few years it has become evident that ecological problems may constrain the use of some technologies.

The link between the introduction of a new technology and the gradual appearance of ecological problems is not new. Some 6,000 years ago, a civilization flourished on the floodplain of the Tigris and Euphrates Rivers, in what is now Iraq, as the development of irrigation technologies turned the desert into fertile land. Gradually, over the course of several centuries, however, the fields became a salty wasteland. Crop yields slowly declined, until production was no longer sufficient to sustain the civilization. The problem was caused by waterlogging of the subsoil and by the constant evaporation of irrigation waters, which left behind dissolved salts. The soil has not yet recovered, and in parts of southern Iraq the earth still glistens with encrusted salt. These particular irrigation technologies were not sustainable over the long term.[59]

The world is not on the verge of ecological collapse, but there is mounting evidence that many technologies being used today are not ecologically sustainable because of their long-term effects on people or on nature. One troubling possibility, for example, is that human activities may lead to irreversible changes in the earth's climate.

When it was first suggested a few years ago that the global climate may eventually show signs of human interference, the idea was greeted with due skepticism by many scientists. Recently, however, the skepticism has given way to concern. It is thought that carbon dioxide—an inevitable by-product of burning fossil fuels—is building up in the atmosphere and acting rather like a greenhouse, preventing a fraction of the earth's heat from being radiated into space. The ultimate result could be a rise in the average global temperature.[60]

The concentration of carbon dioxide in the atmosphere is already believed to have risen by about 13 percent since the Industrial Revolution, and it may double over the next 50 years. Such a change could increase average temperatures on the earth's surface by 2-3°C. The impact of a temperature shift of that magnitude is difficult to predict, but it is likely to affect agriculture in many regions. It could, for example, push the American corn belt northward onto less fertile soils. But it could also extend the growing season in the Soviet grain-producing region and draw monsoon rains into higher latitudes, which would benefit China's rice cultivation. In general, global warming would bring increased rainfall, and probably cause local weather patterns to become more variable.[61]

A more worrying possibility is that global warming could have an adverse impact on the stability of the polar ice caps. According to J. H. Mercer, in an article published in *Nature*, the anticipated doubling of carbon dioxide concentration over the next 50 years could raise polar temperatures by an amount sufficient to cause the West Antarctic ice sheet to break up. Such an event would raise the average sea level by about five meters, which would be catastrophic for many low-lying areas. As Mercer puts it: "If the present highly simplified climatic models are even approximately correct, this deglaciation may be part of the price that must be paid in order to buy enough time for industrial civilization to make the changeover from fossil fuels to other sources of energy. If so, major dislocations in coastal cities, and submergence of low-lying areas such as much of Florida and the Netherlands, lie ahead."[62] Some of Asia's principal ricelands, such as the river floodplains in Bangladesh and Thailand, would also be inundated with salt water.

Although many uncertainties surround the predicted links between carbon dioxide in the atmosphere and global warming, the potential for serious and irreversible climate change provides an additional incentive to push ahead with a major program of energy conservation and development of renewable energy resources. Climatologist Stephen Schneider notes that if we wait until the carbon dioxide greenhouse theory has been proven correct by a warming of the atmosphere, it will be too late to take action. "Ten years of delay will put us ten years closer to potentially irreversible changes," Schneider warns.[63]

While the carbon dioxide greenhouse effect may require major adjustments in energy policies and in long-term planning in many countries, another predicted link between technology and changes in the atmosphere has been easier to resolve. In 1974, researchers at the University of California suggested that fluorocarbon gases, which are widely used in aerosol spray cans, may damage the earth's protective ozone layer. Because the ozone layer helps block some ultraviolet radiation from reaching the earth's surface, it plays a critical role in shielding plants and animals from some of the damaging effects of the sun's rays. Among other things, certain kinds of ultraviolet radiation have been linked with skin cancer.[64]

33

Although the suggested links between fluorocarbons released from spray cans, damage to the ozone layer, and increased numbers of skin cancer victims are not accepted by all scientists, a string of committees, including one convened by the National Academy of Sciences, have reported that the matter warrants serious concern. Consequently, the use of fluorocarbons in spray cans is being phased out through government regulations in the United States and in some European countries. Unlike coal, fluorocarbons fortunately play only a peripheral role in the economies of the industrial countries, and their potential ecological threat has been easily dealt with.[65]

A more difficult ecological problem linked with the introduction of a new technology has arisen in some regions from overuse of pesticides. When DDT was introduced into agriculture in the late forties, it seemed like a miracle cure for a problem that had dogged farmers for centuries. The bollworm, which had decimated U.S. cotton fields, was brought under control in dramatic fashion, for example. But by the mid-fifties, worrisome problems began to emerge.

The budworm, an insect that had not previously been a major cotton pest, developed resistance to DDT and eventually to other pesticides as well. It subsequently assumed a leading role in the destruction of cotton crops in many regions. By the late sixties, farmers in the Rio Grande Valley were desperately spraying 35 times a year in attempts to control the budworm. Many cotton producers eventually went out of business. Similar problems developed in other countries where American techniques were adopted. Cotton farmers in Guatemala and Nicaragua, for example, have increased pesticide applications from 8 a year in the forties to about 40 a year in the mid-seventies. And overuse of pesticides in Peru's fertile Cañete Valley caused serious economic problems in the fifties as extra sprayings to control resistant pests reduced the farmers' incomes.[66]

Cotton is perhaps the most important U.S. crop to develop problems with resistant pests, for it now accounts for nearly half of all the pesticides used there. But it is by no means the only one. The onion maggot, a major pest of New York's onion crop, has developed resistance to a series of pesticides, and only one effective compound is left on the market. Soybeans in the United States are threatened by an insect that is beginning to acquire resistance to the only effective pesticide now in production. And out of 20 pesticides recommended for use in Michigan apple orchards in the past two decades, only 5 are still in use; resistance has developed to the other 15.[67]

In the future, pest control must turn to techniques that are ecologically more sustainable. Biological controls, such as the introduction of natural enemies of crop pests, the release of sterile male insects to reduce breeding, and the use of sex attractants to lure pests into traps are already being tried. Another effective strategy is crop rotation, which reduces losses from pests that survive in the soil during the winter.

Although some farmers are skeptical that such technologies will prevent major infestations, these methods are already widely used in some countries and their use is growing in the United States. A particularly striking example of ecologically sound pest management occurred in the Peruvian Cañete Valley after problems developed with conventional pesticides. In 1957, farmers in the valley organized an areawide control program that included the introduction of enemies of the cotton pests and more resistant cotton varieties, the rotation of crops, and the use of mineral insecticides only when necessary. Synthetic organic pesticides were banned. Production rose dramatically, almost doubling in seven years. A major incentive for farmers to turn to ecologically more sustainable techniques is the fact that overuse of pesticides has become economically unsustainable.[68]

The popular and scientific press abounds with examples of other technologies whose use may be ecologically unsustainable over the long term. Sometimes the problems show up fairly quickly. In a hydro-electric project at Anchicaya, Colombia, for example, the reservoir lost four-fifths of its capacity in just 15 years because of sedimentation, and salinity problems have appeared along the Colorado River as a growing number of irrigation projects return salty water to the river. More often, the ecological effects of a new technology take decades to become apparent.[69]

35

Such is the case with many carcinogens introduced into the environment, particularly in the workplace, through new technologies. The links between asbestos and lung cancer, vinyl chloride and liver cancer, and the drug diethylstilbestrol (DES) and vaginal cancer are but a few that have been definitively established many years after the introduction of each new compound. Each year, about 1,000 new chemical compounds are produced in the United States, yet until the passage in 1976 of the landmark Toxic Substances Control Act, there was no legal requirement that they be tested for long-term toxic effects.[70]

These are just a few examples of technologies whose impact on people or the environment may be unacceptable over the long term. In some cases, such as environmental pollution associated with specific technologies, new technology may alleviate the impacts. In other cases, such as the use of fluorocarbons in aerosol spray cans, the technology itself must be abandoned. Unlike civilizations in the past that have been confronted with potentially catastrophic ecological threats, humanity today at least has the ability to predict some dangers decades in advance. Whether that foresight will lead to corrective action is, however, another matter.

Technological Choices in Context

The relationship between technology and society is a two-way process. Technology has provided much of the driving force behind social change for thousands of years, and it underpins current economic and social systems. But social values, institutions, and political structures shape both the development and adoption of technologies. The pace of social change has quickened in the past few decades as the resources devoted to research and development have grown rapidly, and as institutions for disseminating new technologies have proliferated.

This vast infusion of technology into society has brought immense benefits. But it has not been without social and environmental costs. The four concerns discussed in this paper—rising unemployment, growing social inequities, dwindling oil and gas reserves, and potential long-term ecological problems—are all linked with technology. Some of the technological trends of the past few decades are not compatible with the social needs and resource constraints that lie ahead. Yet choices of technology made today by individuals, communities, corporations, and governments will have lasting impacts on the use of energy and resources. They will affect employment and income distribution for many years to come. Unless consideration of such impacts enters into judgements of which technologies should be developed and employed for particular tasks, some technological choices will lead to more problems than they solve.

Technological choices, whether in industrial or developing countries, are never made in a political or economic vacuum. The entire innovation process, from basic research to the introduction of a new technology, is conditioned by such factors as the profit motive, prestige, national defense needs, and social and economic policies. Those forces must be understood in any discussion of appropriate technology.

If technological development is to be more compatible with human needs, and more in harmony with the earth's resources, four principal points must be recognized. First, the unfettered workings of the market system cannot be relied upon to promote the development and adoption of appropriate technologies. Second, many technologies produced in the past few decades are becoming inappropriate in the industrial countries, and they are even less appropriate to the needs of developing countries. Third, the development of technologies that mesh with local needs and resources requires that developing countries be able to generate and apply new technologies, and it may also require new arrangements for sharing technologies within the Third World. Fourth, it must be accepted that technology, by itself, cannot solve political and social problems.

In most societies, market forces are the principal factor influencing the development and adoption of technologies. But they are at best an imperfect mechanism for ensuring that the development and introduction of new technologies will be socially and environmentally

"Technological choices, whether in
industrial or developing countries,
are never made in a political
or economic vacuum."

acceptable. For one thing, the negative impacts of new technologies
are seldom completely reflected in market prices. A few years ago, for
example, some Canadian pulp and paper manufacturers routinely
flushed mercury wastes down the drain, a process that was much
'cheaper" than installing pollution-control equipment. The costs
were borne by the people who ate contaminated fish rather than by
the manufacturers or the users of pulp and paper products. Government regulations have since banned such socially irresponsible practices, and pollution-control equipment has been installed. Market
prices now better reflect the true costs of the manufacturing process.[71]

37

Market processes can also work to perpetuate the use of inappropriate technologies. American automobile manufacturers find large,
gasoline-guzzling automobiles more profitable to market than smaller,
more energy-efficient cars, for example. In the absence of government regulation, Detroit would pay little heed to fuel efficiency, but
federal regulations are forcing the automobile industry to shift its
production gradually toward more efficient vehicles.

Governments have always strongly influenced trends in technological
development, either directly through research and development and
actual purchases of new technologies, or indirectly through subsidies,
tax incentives, pricing policies, and support for such sectors as education, road building, health care, and so on. Nuclear power, for
example, would not have been developed without immense government support; communications satellites owe their existence to a vast
infrastructure of hardware and expertise that has been developed
through national space programs; and transportation technologies
require government assistance in the form of roads, airports, and
docks. Governments therefore have great leverage in directing technological development along appropriate—or inappropriate—paths.

Most governments now accept some responsibility for ensuring that
costs such as pollution and health hazards associated with some
technologies are borne by the manufacturers and users of those technologies rather than by society at large. There is, however, need for
more systematic methods for anticipating potential side effects of
new technologies. The recent passage in the United States of the
Toxic Substances Control Act, which requires preliminary testing of
new chemical compounds for long-term toxicity, is a step in this

38

direction. The Office of Technology Assessment, established as an agency of the U.S. Congress in 1973, has also published a number of studies highlighting potential long-term problems linked with new technologies. Such mechanisms can help to provide early warnings of side effects of technologies in time to take corrective action. But the development of appropriate technologies, as opposed to the control of inappropriate technologies, requires more direct intervention in market demands.

In the energy field, for example, government subsidies, direct purchases of equipment, and tax incentives could play an important role in promoting the development of renewable energy resources. Two serious impediments to the widespread use of solar equipment are the high immediate cost of the hardware to the consumer (despite its savings over the long run), and the lack of access to long-term credit facilities. Tax incentives, such as the recently adopted 55-percent income tax credit on all solar equipment in California, could play a significant role in stimulating the development of a solar heating industry, while a major government purchasing program could help bring down the cost of photovoltaic cells.[72]

The market system determines the price of goods and services in relation to current supply and demand. But it does not reflect the cost to future generations of the depletion of resources or of environmental degradation. Those costs can only be dealt with by effective conservation of the earth's resources through public policy.

In the developing countries, the market mechanism cannot work to stimulate the development and introduction of technologies that meet the needs of the poor, for the simple reason that the poor, by definition, are often outside the market system. Unless governments, foreign aid agencies, and community organizations assume responsibility for bringing appropriate technologies to subsistence farmers, small-scale manufacturers, and others now outside the market system, the poor may not benefit at all from technological progress.

Until recently, it has been tacitly assumed that Western-style industrial development would be the appropriate model for developing countries to follow. Just as similar technologies are now employed throughout the industrial countries, it was generally anticipated that

"Far from being a technological
monoculture, the world of the future
will have to be characterized
by technological diversity."

he world would eventually be transformed into a sort of technolog-
cal monoculture, with the same agricultural systems, transportation
echnologies, industrial processes, and building techniques used
around the globe. But such assumptions were never valid.

The energy-intensity and materials requirements of many modern
technologies make their use questionable not only in the developing
countries but in the industrial world as well. Moreover, the costs—
both in terms of capital requirements and social impacts of massive
transfers of technology from rich to poor countries—would be pro-
hibitive. Far from being a technological monoculture, the world of
the future will have to be characterized by technological diversity if
it is to be socially and ecologically sustainable. Each society will have
to determine for itself what is appropriate in terms of its own needs
and resources. No two societies are likely to need exactly the same
mix of technologies.

The mechanisms for introducing technologies into developing countries—whether governments, corporations, or international agencies—
all tend to promote a one-way flow from the industrial countries to
the developing world. This is not surprising in view of the fact that
virtually all the world's technological capacity is concentrated in the
affluent nations of the North. But one result is that many technologies do not make effective use of either the physical or the human resources of developing countries. The technologies for making
paper, for example, are based primarily on the use of softwoods that
grow in cool climates; until recently, little attention was paid to the
development of technologies that use cellulose materials abundant
in the tropics. Similarly, until the past few years, small-scale energy
technologies suitable for application in remote villages in the Third
World received scant support from the world's technological powers,
or from the elites in poor countries.

Aside from the fact that technologies developed in the industrial
world may not mesh with the needs of the Third World—or indeed
with the changing social and environmental conditions in the industrial countries themselves—the transfer of technology from rich to
poor countries imposes a heavy financial burden. The developing
countries are now paying about $10-20 billion a year directly and
indirectly for imported technologies, and if present trends continue,
those costs could soar to more than $150 billion by the turn of the

century. Such projections assume, however, that developing countries will continue to buy many technologies that are inappropriate for their needs. Moreover, a group of development experts convened by the United Nations Conference on Trade and Development (UNCTAD) recently pointed out that "it will never be possible to build a world community based on genuine interdependence and capable of tackling global problems as long as three-fourths of that community is dependent on the other fourth for the ability to solve its own problems." The UNCTAD group went on to argue for greater technological self-reliance in the Third World.[73]

There is no easy formula for breaking the technological dependence of the developing world on the advanced industrial countries. But an important element in any such strategy is the strengthening of the capacity of developing countries to meet some of their own technological needs. This does not mean that the Third World should invest scarce resources in research and development institutes that simply copy those in the industrial countries. Rather, the need is to determine indigenous technological requirements and remove barriers to the development and introduction of technologies keyed to local needs and resources.

There is also scope for strengthening the technological links among developing countries. Because many countries face common problems that are not being dealt with adequately by the current technological world order, great potential exists for joint projects, information-sharing, and even transfer of technologies within the developing world. Indian scientists, for example, are now studying closely Chinese biogas plants to see whether they would be suitable for use in Indian villages. A new type of cement made from the ash of rice husks, recently developed in India, could be useful as a partial replacement for building cement in many developing countries. And sustainable, productive, dry-land farming techniques that have evolved in northern Nigeria may be relevant to the needs of farmers in arid zones elsewhere in Africa, Asia, and Latin America.[74]

It should be emphasized, however, that technological changes are only one influence on poverty, unemployment, and other pressing problems in the developing countries. Political, social, and economic transformations are also required to help raise the living standards of those who are now at subsistence level. Charles Weiss, science adviser to the World Bank, has noted that "evidence is piling up that the impact of the introduction of any particular piece of equipment—whether tractors in South Asia or waterless toilets in Viet Nam—depends heavily on the social and institutional structures on which it is superimposed. For this reason, there are many situations in which an intervention focused purely on technology—whether indigenous or foreign, new, adapted, or transferred—is likely to be doomed from the start."[75]

During the past decade, disenchantment with various aspects of the current technological world order has begun to manifest itself in various forms. Developing countries have argued strongly for a restructuring of the global economic system, and their frustrations with the technological dominance of the industrial nations lie behind the plans for a United Nations Conference on Science and Technology for Development scheduled to take place in August 1979 in Vienna. And in the industrial countries, the rise of the environmental movement, the emergence of the technology assessment movement, and the formation of appropriate technology groups signify a measure of disaffection with some aspects of modern technology.

It is relatively easy to identify some of the key criteria for determining whether a technology is appropriate, but far more difficult to devise the social mechanisms to ensure that appropriate technologies are developed and applied. It is also difficult to generalize from one society to another. Only by paying careful attention to the impact of new technologies on people, social systems, and the natural environment will the picture of an appropriate technology for any particular situation begin to emerge. There are, however, no technological panaceas. Only hard choices.

Notes

1. For a discussion of changing attitudes toward science and technology, see Albert H. Teich, ed., *Technology and Man's Future* (New York: St. Martin's Press, 1977) and Stewart L. Udall, "The Failed American Dream," *Washington Post*, June 12, 1977.

2. E. F. Schumacher, *Small Is Beautiful: Economics As If People Mattered* (London: Blond and Briggs, 1973).

3. For a discussion of the social and political dimensions of appropriate technology, see David Dickson, *The Politics of Alternative Technology* (New York: Universe Books, 1975).

4. International Labour Office, "World and Regional Labour Force Prospects to the Year 2000," in *The Population Debate: Dimensions and Perspectives, Papers of the World Population Conference* (New York: United Nations, 1975); International Labour Office, *Employment, Growth and Basic Needs: A One-World Problem* (New York: Praeger Publishers, for the Overseas Development Council, 1977).

5. There are few good estimates of the capital costs of providing jobs in industry, but the costs do not differ much in modern industries established in developing or industrial countries. The Conference Board has estimated that the capital investment per job in the United States ranges from $108,000 per employee in petrochemicals industries to $5,000 in the clothing industry, with an average of about $20,000 in all industries. See Richard Grossman and Gail Daneker, *Jobs and Energy* (Washington, D.C.: Environmentalists for Full Employment, 1977).

6. A good general framework for the economics of appropriate technology is Hans Singer, *Technologies for Basic Needs* (Geneva: International Labour Office, 1977) and Frances Stewart, *Technology and Underdevelopment* (Boulder, Col.: Westview Press, 1977).

7. World Bank, "Appropriate Technology in World Bank Activities," Washington, D.C., July 1976.

8. John P. McInerney *et al.*, "The Consequences of Farm Tractors in Pakistan," World Bank, Washington, D.C., 1975.

9. A good general discussion of the peak labor demand problem is Arjun Makhijani and Alan Poole, *Energy and Agriculture in the Third World* (Cambridge, Mass.: Ballinger Publishing Co., 1975). Stuart S. Wilson, "Pedal Power," in R. J. Congdon, ed., *Introduction to Appropriate Technology* (Emmaus, Penn.: Rodale Press, 1977); Amir U. Khan, "Mechanization Technology for Tropical Agriculture," in Nicholas Jequier, ed., *Appropriate Technology: Problems and Promises* (Paris: Organization for Economic Cooperation and Development, 1976); World Bank, "Appropriate Technology."

10. Denis N. Fernando, "Low-Cost Tube Wells," *Appropriate Technology*, Vol. 2, No. 4, 1976; World Bank, "Appropriate Technology"; Peter Fraenkel, "Food from Wind in Ethiopia," *Appropriate Technology*, Vol. 2, No. 4, 1976.

11. Iftikhar Ahmed, "Appropriate Rice Production Technology for Bangladesh," *Agricultural Mechanization in Asia*, Autumn 1977.

12. World Bank, *Appropriate Technology and World Bank Assistance to the Poor*, S & T Report No. 29, Washington, D.C., December 1977; M. Allal et al., "Development and Promotion of Appropriate Road Construction Technology," *International Labour Review*, September/October 1977.

13. American Rural Small-Scale Industry Delegation, *Rural Small-Scale Industry in the People's Republic of China* (Berkeley: University of California Press, 1977).

14. Jon Sigurdson, "Rural Industrialization in China: Approaches and Results," *World Development*, July/August 1975.

15. American Rural Small-Scale Industry Delegation, *Rural Small-Scale Industry in China*.

16. For a general discussion of the role of small-scale industries in development, see David Vail, "The Case for Rural Industry," Program on Policies for Science and Technology in Developing Nations, Cornell University, Ithaca, N.Y., July 1975; for a discussion of Taiwan, see Keith Griffin, *Land Concentration and Rural Poverty* (New York: Holmes and Meier, 1976). A central feature of Taiwan's development strategy is the establishment of many small market towns that serve as agro-industrial centers. These ideas are discussed in Edgar Owens and Robert Shaw, *Development Reconsidered* (Lexington, Mass.: D.C. Heath & Co., 1972).

17. India's industrial development is analyzed in John W. Mellor, *The New Economics of Growth* (Ithaca, N.Y.: Cornell University Press, 1976).

18. "India's De-Industrial Revolution," *The Economist*, November 19, 1977.

19. M. K. Garg, "The Scaling-Down of Modern Technology: Crystal Sugar Manufacturing in India," in Jequier, *Appropriate Technology*.

20. An excellent discussion of the institutional barriers to the introduction of appropriate technology in developing countries is given by Nicholas Jequier in his introduction to *Appropriate Technology*.

21. "World Job Losses Put at 40-Year High," *New York Times*, November 30, 1975; "Canada's Unemployment of 8.5% Is Highest Since Depression," *New York Times*, January 11, 1978; "European Unemployment Goes Over 6 Million," *Christian Science Monitor*, January 11, 1978.

22. Lester Thurow, "Inequality, Inflation, and Growth in the American Economy," *The Economist*, December 24, 1977.

23. "White House Says Weather-Crisis Layoffs Totaled 1.8 Million at Its Peak," *New York Times*, February 8, 1977.

24. Denis Hayes, *Rays of Hope: The Transition to a Post-Petroleum World* (New York: W. W. Norton & Co., 1977).

25. Fred Branfman and Steve LaMar, *Jobs From the Sun* (Los Angeles: California Public Policy Center, February 1978); Hugh Fitzpatrick, "A Comparative Analysis of the Employment Effects of Solar Energy in California," Employment Development Department, Sacramento, Calif., 1978.

26. A discussion of the link between income distribution and technology choice is given in "A Conceptual Framework for Environmentally Sound and Appropriate Technologies," report of United Nations Environment Programme Expert Group Meeting held in Nairobi, Kenya, December 1-4, 1975.

27. World Bank, "Appropriate Technology for Water Supply and Waste Disposal," Progress Report, Washington, D.C., June 1977.

28. Keith Griffin, *The Green Revolution: An Economic Analysis* (Geneva: United Nations Research Institute for Social Development, 1972); Nicholas Wade, "The Green Revolution: A Just Technology, Often Unjust in Use," *Science*, December 20 and 27, 1974.

29. International Labour Office, *Towards Full Employment: A Programme for Colombia* (Geneva: 1970); Makhijani and Poole, *Energy and Agriculture in the Third World*.

30. D. C. Rao, "Urban Target Groups," in Hollis Chenery, ed., *Redistribution With Growth* (London: Oxford University Press, 1974); World Bank, *Global Estimates for Meeting Basic Needs: Background Paper* (Washington, D.C.: Policy Planning and Program Review Department, August 1977).

31. Charles Elliott, *Patterns of Poverty in the Third World* (New York: Praeger Publishers, 1975).

32. Richard Jolly, "International Dimensions," in Chenery, *Redistribution With Growth*.

33. Bruce Stokes, *Local Responses to Global Problems: A Key to Meeting Basic Human Needs* (Washington, D.C.: Worldwatch Institute, February 1978); information on Sri Lanka from *The United States and World Development: Agenda 1977* (New York: Praeger Publishers, for the Overseas Development Council, 1977).

34. Mahbub ul Haq, *The Poverty Curtain: Choices for the Third World* (New York: Columbia University Press, 1976).

35. Lester R. Brown, *The Twenty-Ninth Day: Accommodating Human Needs and Numbers to the Earth's Resources* (New York: W. W. Norton & Co., 1978).

36. Frank Melville, "The Concorde's Disastrous Economics," *Fortune*, January 30, 1978.

37. C. R. Prasad, K. Krishna Prasad and A.K.N. Reddy, "Biogas Plants: Prospects, Problems and Tasks," *Economic and Political Weekly*, August 1974.

38. See, for example, James Howe *et al.*, *Energy for Developing Countries* (Washington, D.C.: Overseas Development Council, forthcoming) and Denis Hayes, *Energy for Development: Third World Options* (Washington, D.C.: Worldwatch Institute, December 1977).

39. Vaclav Smil, "Energy Solution in China," *Environment*, October 1977; Joe Whitney, "Fueling China's Development," *New Internationalist*, March 1978.

40. Prasad *et al.*, "Biogas Plants."

41. Estimates of global oil reserves are a topic of intense debate, but most projections indicate that a downturn in production will take place around the year 2000. In particular, see the report of the Workshop on Alternative Energy Strategies, *Energy: Global Prospects 1985-2000* (New York: McGraw-Hill, 1977) and Peter Nulty, "When We'll Start Running out of Oil," *Fortune*, October 1977. For a discussion of nuclear problems see Denis Hayes, *Nuclear Power: The Fifth Horseman* (Washington, D.C.: Worldwatch Institute, May 1976) and William Metz, "Nuclear Goes Broke," *The New Republic*, February 25, 1978. The link between carbon dioxide and climate is analyzed in National Academy of Sciences, *Energy and Climate* (Washington, D.C.: 1977).

42. Carol and John Steinhart, *Energy: Sources, Use, and Role in Human Affairs* (North Scituate, Mass.: Duxbury Press, 1974).

43. David Pimentel *et al.*, "Food Production and the Energy Crisis," *Science*, November 2, 1973; David and Marcia Pimentel, "Counting the Kilocalories," *Ceres*, September/October 1977.

44. Pimentel, "Food Production"; Steinhart, *Energy in Human Affairs*.

45. Pimentel, "Counting the Kilocalories."

46. *Ibid.*; A.K.N. Reddy and K. Krishna Prasad, "Technological Alternatives and the Indian Energy Crisis," *Economic and Political Weekly*, August 1977.

47. For a discussion of energy and fertilizers, see Brian Pain and Richard Phipps, "The Energy to Grow Maize," *New Scientist*, May 15, 1975.

48. Jerome Goldstein, *Sensible Sludge* (Emmaus, Penn.: Rodale Press, 1977). In the United States, the Environmental Protection Agency now requires all local authorities to look into the possibility of recycling sludge on farmland as an alternative to building new treatment plants, as reported in "EPA Shifts to Emphasize Recycling Sewage on Land," *Washington Post*, October 13, 1977.

49. Erik Eckholm, *The Other Energy Crisis: Firewood* (Washington, D.C.: Worldwatch Institute, September 1975).

50. Arjun Makhijani, "Solar Energy and Rural Development for the Third World," *Bulletin of the Atomic Scientists*, June 1976; Ianito Evans and Donald Wharton, "The Lorena Mudstove: A Wood-Conserving Cookstove," *Appropriate Technology*, Vol. 4, No. 2, 1978.

51. Reddy and Prasad, "Technological Alternatives."

52. Eric Hirst, "Energy Intensiveness for Transportation in the U.S.," Oak Ridge National Laboratory, Oak Ridge, Tenn., April 1973, as cited in Wilson Clark, *Energy For Survival* (New York: Anchor Books, 1974).

53. Clark, *Energy for Survival.*

54. Stan Luxenberg, "Bicycle Makers are Getting Back on the Track," *New York Times*, January 29, 1978; Andrew H. Malcolm, "Japan Cycles into a New Problem," *New York Times*, April 8, 1978; "Bicycles," *TRANET*, Fall 1977; *The Davis Experiment* (Washington, D.C.: Public Resource Center, 1977); James C. McCullagh, ed., *Pedal Power* (Emmaus, Penn.: Rodale Press, 1977).

55. American Institute of Architects, *Energy and the Built Environment* (Washington, D.C.: 1975).

56. Environmental Protection Agency, *Resource Recovery and Waste Reduction, Fourth Report to Congress* (Washington, D.C.: 1977).

57. Hayes, *Rays of Hope.*

58. Amory B. Lovins, *Soft Energy Paths: Toward a Durable Peace* (Cambridge, Mass.: Ballinger Publishing Co., 1977).

59. Erik Eckholm, *Losing Ground: Environmental Stress and World Food Prospects* (New York: W. W. Norton & Co., 1976).

60. National Academy of Sciences, *Energy and Climate.*

61. Charles F. Cooper, "What Might Man-Induced Climate Change Mean?," *Foreign Affairs*, April 1978; for a general account of climate change, see Stephen Schneider with Lynne Mesirow, *The Genesis Strategy: Climate and Global Survival* (New York: Plenum Press, 1976).

62. J. H. Mercer, "West Antarctic Ice Sheet and CO_2 Greenhouse Effect: A Threat of Disaster," *Nature*, January 26, 1978.

63. Stephen Schneider, "Climatic Limits to Growth: How Soon? How Serious?," presented to the 144th Annual Meeting of the American Association for the Advancement of Science, Washington, D.C., February 17, 1978.

64. M. J. Molina and F. S. Rowland, "Stratospheric Sink for Chlorofluoromethanes: Chlorine Atom-Catalysed Destruction of Ozone," *Nature*, June 28, 1974.

65. National Academy of Sciences, *Halocarbons: Effects on Stratospheric Ozone* (Washington, D.C.: September 1976).

66. Daniel Zwerdling, "The Pesticides Plague," *Washington Post*, March 5, 1978; Erik Eckholm and S. Jacob Scherr, "Double Standards and the Pesticide Trade," *New Scientist*, February 16, 1978; Teodoro Boza Barducci, "Ecological Consequences of Pesticides Used for the Control of Cotton Insects in Cañete Valley, Peru," in M. Taghi Farvar and John P. Milton, eds., *The Careless Technology* (New York: Natural History Press, 1972).

67. Zwerdling, "The Pesticides Plague." National Academy of Sciences, *Contemporary Pest Control Practices and Prospects* (Washington, D.C.: February 1976) provides an excellent discussion of pest control problems in the United States.

68. Paul Shinoff, "Big Farms Adopt Organic Methods to Control Pests," *Washington Post*, January 9, 1978; Barducci, "Ecological Consequences of Pesticides."

69. Penny Lernoux, "Ecological Disaster Threatens Colombia's Hydroelectric Projects," *World Environment Report*, December 5, 1977.

70. For a discussion of environmental sources of disease, see Erik Eckholm *The Picture of Health: Environmental Sources of Disease* (New York W. W. Norton & Co., 1977).

71. Science Council of Canada, *Canada as a Conserver Society* (Ottawa September 1977).

72. The role of tax incentives in supporting solar development is discussed by Henry Kelly in his statement before the Subcommittee on Domestic Monetary Policy, U.S. Senate, Committee on Banking, Finance, and Urban Affairs, April 25, 1978.

73. Carlos Contreras *et al.*, "Technological Transformation of Developing Countries," based on a meeting held in Geneva, January 1978, and published as *Discussion Paper No. 115* by Research Policy Program, University of Lund, Sweden, February 1978.

74. These issues will be discussed at the United Nations Conference on Technical Cooperation Between Developing Countries in Buenos Aires Argentina, in August 1978. "India Adopts Chinese Biogas Designs," *Development Forum*, March 1978; Anil Agarwal, "Can Industry Grow Without Agriculture?," *New Scientist*, April 20, 1978; Erik Eckholm and Lester R Brown, *Spreading Deserts—The Hand of Man* (Washington D.C.: Worldwatch Institute, August 1977).

75. Charles Weiss, "Technology in Context," presented to the Workshop on Scientific and Technological Cooperation With Developing Countries, Organization for Economic Cooperation and Development, Paris, April 10-13, 1978.

COLIN NORMAN is a Senior Researcher with Worldwatch Institute. His research deals with the social and political issues connected with the choice and control of technology. Prior to joining Worldwatch, he was the Washington correspondent for *Nature*.